REFUSE TRUCKS

PHOTO ARCHIVE

John B. Montville

Iconografix
Photo Archive Series

Iconografix
PO Box 446
Hudson, Wisconsin 54016 USA

Library of Congress Card Number: 00-135945

ISBN 1-58388-042-9

01 02 03 04 05 06 07 5 4 3 2 1

Printed in the United States of America

Cover and book design by Shawn Glidden

Editing by Dylan Frautschi

COVER PHOTO:(See page 71)

Book Proposals

Iconografix is a publishing company specializing in books for transportation enthusiasts. We publish in a number of different areas, including Automobiles, Auto Racing, Buses, Construction Equipment, Emergency Equipment, Farming Equipment, Railroads & Trucks. The Iconografix imprint is constantly growing and expanding into new subject areas.

Authors, editors, and knowledgeable enthusiasts in the field of transportation history are invited to contact the Editorial Department at Iconografix, Inc., PO Box 446, Hudson, WI 54016.

ACKNOWLEDGMENTS

No book of this nature could ever go from conception to finished product without the help of others, and this one is no exception. At the start, Iconografix, Inc., was vital to the selection of a subject regarding the historical use of trucks that has been overlooked. On a subject that took longer to illustrate properly than first envisioned, I must show gratitude to all those whose aid and patience were vital to its successful completion.

My good friend, Al Velocci of Long Island, New York, was most supportive of this project. As a former owner of a private sanitation company, his knowledge of the industry and referrals therein were most helpful. Also, as a devotee to automotive and local history, Al shared some historical information that was most helpful to my understanding the development of the industry.

My friends Stu Maguire and Andy Hill, of upstate New York, were most helpful with information and graphics on certain postwar trucks. My contact with Stu goes back to 1951 and my first correspondence with him as a youthful heavy-equipment fan. He loaned some important photos and truck body literature from the 1950s, which helped to cover this important era. Andy Hill is doing his best to keep the "Brockway" name alive through his Brockway Trucks National Registry, and publishes "Brockway Today" on a quarterly basis. His contact with the Cortland Historical Society was most important for the coverage of Brockway contributions to the postwar refuse-collecting industry.

Another long-time friend, Bill West, was also kind in providing help in covering some West Coast developments, and supplied photos of Fageol refuse trucks from the Bay Area. Another West Coast truck enthusiast and truck book author, Don Wood, supplied information and a graphic related to DeMartini scavenger trucks. Another friend, George Fiebe, a long-time truck enthusiast with many over-the-road miles behind him, helped with both specific items and general information on the subject. Fellow truck enthusiast and Walter truck history author, Mark Simiele, helped with a photo and information on Walter trucks.

No book covering truck bodies would be complete without some input from Charles Wacker, former Philadelphia body builder. Charles supplied information on Philadelphia's use of certain refuse equipment as well as their builders. Another former truck-industry participant, Gene Windt, of the Tucson area, provided vital help with some personal graphics of more modern Pacific Coast refuse trucks. He also helped in the vital area of front-loaders from the early 1980s. Also, Jim Neal of the R. E. Olds Transportation Museum helped with the view of the Wesco-Jet chassis and information on the related Reo involvement. Matt Lee, fire apparatus historian, helped with FWD refuse truck views and information related to their use. Kerry Day provided some early Mack graphics, as well as other refuse equipment information.

While it may be hard to judge which photos are the most important, since all views were vital to showing the overall development of the equipment, the earliest ones are quite rare. In this regard, a word of sincere appreciation must be said for the Municipal Archives of New York City and its director, Mr. Kenneth Cobb. Without these faithfully preserved archival photos, the wide variety of World War I refuse truck photos would have been impossible to access for this presentation.

In regard to industry sources for graphics, two firms stood out for their cooperation in supplying specific views of equipment: Heil Environmental Industries Ltd. and McClain Industries. Mr. Gary Gengozian, Communications Manager at Heil, supplied both vintage and current equipment views of Heil-bodied refuse trucks. Mr. Steve Drexel, Vice President at McClain and E-Z Pack, supplied some views related to prewar Galion and postwar E-Z Pack refuse equipment. It was hoped that other companies in the refuse equipment industry would participate, but a lack of response limited the overall coverage of body builders.

It should be added that the majority of the photos used in this pictorial overview came from the author's own archives of historical truck material collected since the World War II period.

And, finally, to all those not mentioned who also offered aid and a good word or other contribution, a sincere "thank you."

John B. Montville

INTRODUCTION

It should be obvious to most observers of urban and suburban life that the once mundane and often put-down garbage truck has evolved into a technical marvel. This metamorphosis from a basically solid-tired open dump truck of the World War I period to the cost-efficient automated collection vehicle of today is shown in this pictorial review. A basic chronological presentation has been followed to show this development in as objective a way as possible. In this manner the overall trends in both truck and body designs can be followed in each era, without showing favoritism to any one truck or body manufacturer.

The need for civilized societies to dispose of its waste materials has been a problem since the earliest urban communities of Biblical times. Even in Colonial America, city fathers arranged with various individuals to collect household garbage and rubbish. The term "cartmen" was applied to these early-day refuse collectors, who also provided a scavenger service by picking through household rubbish for reusable materials. These refuse haulers were most often under contract from municipalities, but fluctuating salvage values of reusables often played a negative part in the effectiveness of their services. Street cleaning was always a problem, although the tons of animal manure deposited on the streets of many cities usually had a ready use as fertilizer. The evolution of street cleaning as a major municipal duty, which also included the collection of household garbage and rubbish, was a natural one, as all too often domestic refuse would be left along the curb or actually tossed into the street.

The ultimate disposal of unsalvageable items was traditionally in landfills or garbage dumps, but as cities grew, such sites became more restricted. New York City, as early as the 1850s, loaded unsalvageable materials on scows for dumping at the sea, which then brought its own problems. A basic system of separating household garbage (usually made up of vegetable and animal wastes) from rubbish (made up of solid items like bottles, newspapers, rags, etc.) helped in processing the refuse. The collection of ashes from coal furnaces was no basic problem, as they were often mixed with solid rubbish items for continued landfill purposes. Garbage was then rendered in reduction plants in some cities to recover saleable oils used in industry, but economic changes in the 1920s forced cities to think of straight incineration for reducing the volume of disposable wastes.

It was only natural that with the dawn of the auto-age, street-cleaning departments would soon enlist the aid of commercial motor vehicles for more efficiency in their refuse collections. At first, private truckers were hired for specific collection purposes, and demonstrations were also made by motor truck dealers and manufacturers. Finally, by the World War I period, motor trucks and related special equipment had become so relatively dependable and cost-effective, that the substitution of motor-power for horsepower became a reality.

The typical refuse truck of the 1920s was a medium- to heavy-duty chassis with dump-type body, which could hold wet garbage without leaking. While efforts were made to enclose some bodies with side- and top-hinged doors and covers, most refuse bodies built during the 1920s had open tops. However, by the late 1920s mechanical improvements included under-body hydraulic hoists and one line of power-actuated side-loaders.

During the early 1930s, several truck builders introduced special drop-frame models to reduce the body heights for easier loading. Trends during the 1930s included more enclosed bodies with some being arch-top type and cylindrical designs. Also devised by the late 1930s was the packer-type body, which increased the payload capacity by at least 100 percent. Unfortunately, the Great Depression helped slash the operating budgets of most American cities, which restricted their purchase of the more costly improved collection equipment. It was not until the late 1930s that more cities were able to order refuse trucks with fully enclosed bodies with some having escalator-type loading devices. However, many cities stuck with open-bodied dump trucks until after World War II.

Although interrupted by World War II, the late 1940s saw the concept of the packer-type body gain recognition for its basic efficiency. During the decade of the 1950s, various body builders produced many types of bodies, including rear and side bucket loaders, as well as escalator loaders. However, packer-type devices were being adopted by various body builders as part of different body designs to increase payloads.

Major changes took place in refuse collection equipment concepts during the 1960s, including the perfection of front-loaders and roll-off containers. The use of containers in garbage and rubbish collection resulted in a significant reduction in labor costs as well as the potential for worker injury. Also, constant improvements were made in packer-type bodies to increase their size, strength, and compaction pressures for increased payloads.

During the 1970 to 1985 period, packer-type side-loading refuse bodies were developed along with drop-frame truck models to facilitate their use. The concept of dual left- and right-hand control was also introduced as an aid to maximize the utility of side-loading refuse trucks. Low-entry cabs on specialized refuse truck models was also a major addition in the refuse collection field. Experiments with automated loading devices also took place during this period, with the constant idea of lowering collection costs.

Current improvements in collection equipment include the perfection of the automated side-loader, dual-control one-man trucks, and special recycling trucks with compartmented bodies. The refuse collection equipment industry is a very dynamic one with many new concepts and improvements ready to move off the drawing boards and into a street near you!

John B. Montville

A Gramm 6-ton dump truck being demonstrated for rubbish collection in 1913. The New York City agent for the Gramm Motor Truck Company of Lima, Ohio, was C. P. Hexter. Mr. Hexter's demonstration of this unit in the spring of 1913 was an exercise in futility for several reasons. Mainly, the Gramm line of heavy trucks was phased out during 1913 in favor of the new "Willys Utility" delivery truck, due to the Gramm firm's purchase by John N. Willys, of the Willys-Overland Company, Toledo, Ohio.

A rear view of the Gramm demonstrator indicates it is an "automatic dump" truck. This model was originally designed as a contractor's truck, to be used in the hauling of sand and gravel for road construction. This heavy-duty truck's application to municipal refuse collection was dubious at best due to its obvious moderate capacity for its weight, and the extreme height of the body for loading purposes. Also, specialized garbage bodies for heavy trucks still had to be developed at this point.

A 1915 Couple-Gear front-drive electric garbage cart with sump-type body of about 2-yard capacity. The Couple-Gear Freight Wheel Company of Grand Rapids, Michigan, built a front-drive power unit that was used to motorize heavy horse-drawn wagons, often creating a hybrid vehicle of limited utility. This garbage cart was similar in body design to the horse-drawn collection wagon used in New York City at the time.

A 3-wheeled Knox-Martin tractor and dump trailer shown at a relay station in New York City about 1915. The Knox-Martin was a gasoline-engined power unit, or "tractor," that was also used to motorize horse-drawn wagons. This relay station is the early equivalent of the current transfer station, and served the same function in consolidating refuse collections for shipment to a final disposal site. The Knox-Martin tractor became a 4-wheeled design in 1915, and was built in Springfield, Massachusetts, until the early 1920s.

A 1915 G.V. 10-ton gas-electric tractor and semi-trailer with removable containers. The General Vehicle Company of Long Island City, New York, built 12 of these custom power units for refuse collection in New York City. The removable container system was a product of the Charles Hvass Company, also of New York City. Each of the tractor-trailer units were calculated to replace five of the horse-drawn garbage carts then in use. The G.V. firm was a pioneer builder of battery-powered trucks, but went out of business during 1918.

The G.V. tractor-trailer unit with a second deck of containers holding a rubbish collection. In the proposed two-deck container system, ashes or garbage would be loaded into the lower containers and general rubbish, including paper in the upper ones. When the containers had been filled, the unit would be driven to a dock along the East River where a traveling crane system would unload the containers into waiting scows. Tugboats then towed the scows to a processing center for reduction of the basic garbage and the salvage of items sorted from the rubbish.

The Mack Model AC, introduced in 1916, was soon adapted to dump truck service as this 1917 photo indicates. This demonstrator has the Mack combination chain and cable hoist for tipping the body. It also has the special cab construction for use with the Mack hoist mechanism. The all-metal body was fully enclosed with side doors and top hatches. While the truck has a New York license plate, the photo was actually taken in Philadelphia, where the truck was sent for demonstration purposes.

The same Mack refuse truck being loaded with coal ashes along a residential street. All doors have been raised along the curbside for loading by a three-man team during a test run. An order for 32 of the 5 1/2-ton Macks was placed by Philadelphia after the demonstration in 1917. It was a common practice for many years to collect coal ashes separately, as they packed down well to make excellent landfill. Also, by their very nature they were sanitary and often used as fill in public places, such as walks and driveways.

Another view of the Model AC Mack refuse truck, but backed up to a horse-drawn garbage cart for a contrast in size. It can be noted that while they both show similar overall lengths, the Mack had greater payload capacity for its length. However, many streets in the Center City were so narrow that Philadelphia continued to use horse-drawn refuse collection units on a limited basis up to the early 1950s! Also, this version of the AC chassis had the curved rear spring horns common to later AC Mack dump trucks.

Front view of the Model AC refuse truck gives a good idea of the width and lowness of the body. The all-steel body was built by the Charles Hvass Company, which also supplied various types of special equipment for Mack trucks during the 1914-1917 period. During 1917, the Model AC's second year of production, a louvered hood replaced the screened one shown on this unit. The New York "Dealer" license plate indicates the truck was driven to Philadelphia from Mack's Manhattan main service station in New York City.

This photo shows the demonstrator dumping a basic ash collection into a scow at a dock. The body has been raised to its full extent for complete discharge of payload, as local officials, truck company representatives, and sanitation workers stand by. The Mack cable hoist was not offered after 1917, as other types of hoists became available through local truck equipment dealers.

This refuse truck is dumping coal ashes and household rubbish at a landfill in a suburban area. Many cities used solid types of waste materials to raise the level of marshy areas for their eventual placement on the tax rolls as productive land. Household garbage and other types of organic waste were usually brought to pig farms where such existed within reasonable distances from an urban area.

Another late 1916 or early 1917 Model AC Mack with a dump body that has a more traditional tailgate arrangement. This style of body was very adaptable to the collection of ashes and general rubbish. It was also useful for hauling snow in the winter when the city street-cleaning department needed extra dump trucks to dispose of a heavy snowfall that hindered traffic.

This 1917 Mack AC has a deep sump-type dump body for the removal of wet sludge from catch basins located at city intersections. A new style hood, having louvers, was applied to the Model AC early in 1917. In this same year the Model AC also received its famous "Bulldog" nickname. This style of watertight body became very popular for rubbish and general refuse collection in various sizes. Also, the sump-type was often called a "duck-tail" or a "scow" body.

A circa-1920 Bulldog Mack Model AC tractor pulling two side-dump four-wheeled trailers. Several cities in New York State and Ohio used this method of refuse removal. Trailers were usually pulled by horses on their collection routes until filled, at which point they would meet the tractor at a pre-determined point to drop off the full trailer and pick up an empty one. The destinations of these trailer trains were garbage dumps or pig farms several miles outside of the cities.

An early 1920s DeMartini 4-ton model with 15-yard rubbish body at a garbage dump in Alameda, California. The DeMartini became the legendary refuse collection truck of San Francisco and the East Bay area during the 1920s. Refuse was collected for many years by an association of private refuse haulers, called "scavengers," for their system of salvaging any re-usable materials. They favored a very high-sided wooden body and the DeMartini truck chassis, both of which were manufactured locally. The DeMartini truck was custom-built up to at least the early 1930s.

This 1925 4-cylinder, engine-under-seat Autocar has a 16-yard rubbish body. The body was built by the Darien Body Works in South Philadelphia, mainly for the collection of both ashes and rubbish. The heavy-duty dual tailgates were hinged vertically and designed to form as close of a seal as possible for retaining possible liquids in any waste collections. Philadelphia ordered 30 of the units from the Autocar Company of nearby Ardmore.

A 1927 Fageol low-frame chassis with open cab and wooden dump body. Fageol Motors of Oakland, California, was famous during the 1920s for the introduction of the "Safety Coach," with low-slung frame. No doubt Fageol used its bus-building experience to design this special chassis, which has Gruss shock absorbers at the front. This high-sided body was obviously built for a San Francisco Bay Area scavenger service.

A rear view of this 1927 Fageol scavenger truck shows large barn-type doors at the rear of the body. The size of the body may have necessitated a two-man team inside the body to sort and pack the refuse as it was handed in containers through the side door at the front of the body. Folding steps were located just below the side door under the body. The inside crew filled the body from the back to the front, as they separated items of any resale value.

This circa 1928 Fageol refuse truck has a unique, low-sided, all-steel body for the collection of garbage in the East Bay county of Contra Costa, California. The 6-cylinder Fageol, with attachment-3rd axle, has a gross vehicle weight of close to 10 tons. The metal frames over the body were used to support a tarpaulin, which was secured in stages as the body was loaded. The ultimate destination of the refuse collection is indicated in the name of the truck's owners.

A 1928 Mack Bulldog with 7-yard Heil No. 81-A garbage body. This style of sump, or "duck-tail," body was very popular with private refuse haulers in the New York City area for many years. This basic watertight construction made it adaptable for carrying a wide variety of materials, wet or dry. The body was raised by a model No. 6 Heil twin-cylinder hydraulic hoist, which had passed a factory 23-ton load test.

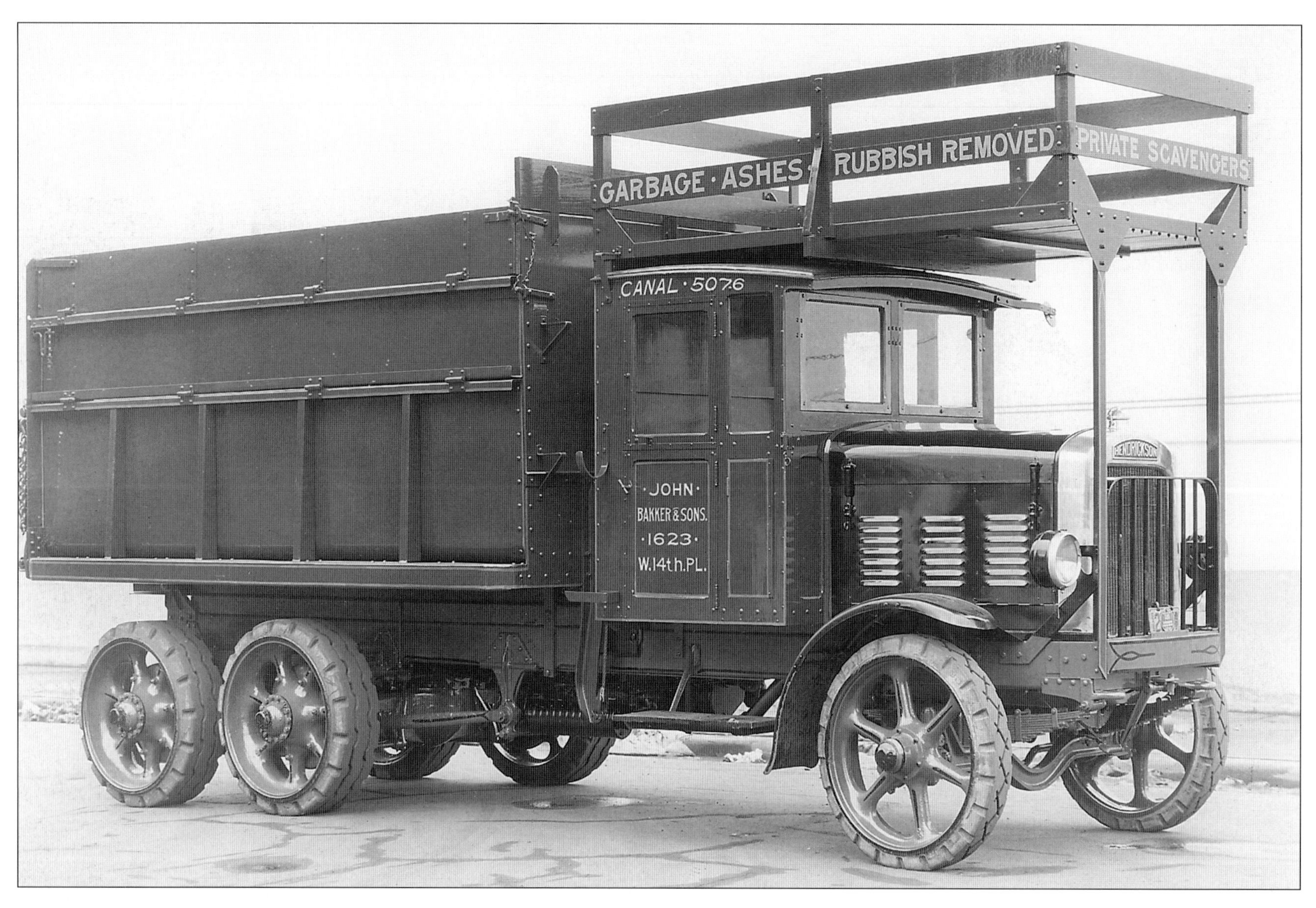

A 1928 Hendrickson heavy-duty 6-wheeler with dump body and cab roof rack. The tandem designation, "6-4-4" meant: six-wheel truck with tandem unit having four-wheel drive and four-wheel brakes. Many Chicago-area private sanitation services favored heavy-duty trucks due to the increasing volume and bulky nature of the material they collected. For many years the Hendrickson Motor Truck Company of Chicago provided a wide variety of custom-engineered trucks to these local scavenger and garbage services.

A 1929 Mack Bulldog with 3-yard Wood garbage body and horizontal hoist. The under-body hoist used a "rolling-wedge" action, which pressed against cams in the lifting mechanism. Also, this truck had fold-down steps below the body to help loaders gain height when lifting containers for emptying. When not in use, the steps folded neatly out of the way.

A side-loading Colecto body on a 1929 heavy-duty American LaFrance truck chassis. The Colecto refuse body was built at this time by the Heil Company for Colecto's New York sales agent, B. Nicol & Co. This body design was an advance in both the efficiency and sanitary aspects of refuse collection. The oblong bucket located along the right side of the body was suspended by cables. After loading, the bucket was raised up the side of the body by a power take-off acting on the cables. The steel apron at the top of the body opened and closed automatically during the dumping process, thus keeping unpleasant odors from escaping.

A left-hand side view of this 1929 Colecto-bodied truck shows a heavy-duty top-hinged tailgate. During discharge the body is lifted to a 45-degree angle by a Heil hydraulic hoist and the tailgate is raised out of the way. This American LaFrance also has front brackets and a control device for the attachment and operation of a snowplow in the winter. Many cities used refuse trucks in dual service to make more efficient use of expensive municipal equipment.

Shown are three Model AB Macks with 15-yard Heil-built "Colecto" bodies owned by Albany, New York. The first truck in this 1929 view has its bucket deployed for loading, while the other two have their buckets raised to the dumping position. Although an improvement in garbage collection bodies, the higher cost of such mechanical equipment restricted their sale during the Depression years of the early 1930s.

The 15-yard Colecto bodies on the Model AB Macks are seen raised to their discharge position. The loading buckets have also been raised to their dump position, which was their normal travel position between collection runs.

A 1930 Ford Model AA with 3-yard Wood open scow-type garbage body. The canvas cover for the body has been removed so that the truck can discharge its load at the transfer station. Three rings along each side of the body were employed with tie ropes to secure the body's cover. The obvious disadvantage of the open-type refuse body was the need to keep the load covered, especially in windy weather.

Shown is the 1930 Ford Model AA with body raised at the transfer station for discharge into an open-top railroad hopper car. A fleet of 100 of these trucks was ordered by the city of Detroit, with Ford providing a 131 1/2-inch wheelbase truck chassis for this particular application.

A 1931 FWD custom-designed drop-frame refuse collection chassis with Wood fully enclosed sanitary body. The Wood single-piston, telescoping, hydraulic hoist (seen at close view) nestled into the front of the large side-loading body. In 1932 the Four Wheel Drive Company introduced the Model LBU, which incorporated front-wheel drive in a drop-frame chassis. During 1931 and 1932 several other custom truck builders introduced special low-frame chassis designed for the garbage and refuse collection industry.

The 1931 Mack prototype drop-frame refuse chassis utilized the Bulldog-style hood and radiator placement. The fully enclosed duck-tail 8-yard body could be tilted to a 50-degree angle to fully discharge its load. Top doors, or roof hatches, could be opened for loading snow in the winter. The solid tires that helped reduce the chassis height were still favored by many heavy truck operators in the New York City area at this time.

A rear axle view of a 1931 Mack drop-frame refuse chassis. The dual reduction Mack axle was held in alignment by tubular radius rods, and axle shafts extended through spaces in the reinforced chassis side-rails. This chassis design avoided the usual high kick-up in the frame to clear the rear axle area found in other drop-frame designs at the time. However, this was basically a prototype chassis, with changes made in the later production models.

A 1932 Model AK Mack drop-frame chassis with Von Keller rotary "Refuse Collector" body. This up-graded prototype has the AK model covered cab, fully crowned fenders, and pneumatic tires for a more up-to-date look. The Von Keller body contained a fixed internal vane, which trimmed the load as the body turned slowly. When the body was ready for emptying, it was elevated and the load was discharged by reverse rotation.

In 1932 the Twin Coach Company of Kent, Ohio, built this prototype 15-yard "Refuse Collector" on a special low-frame chassis. The body was fully enclosed and raised for dumping by a Wood hydraulic telescopic hoist. Four lightweight sliding steel doors were located on each side of the body. The fully enclosed cab had a bench seat in back for a five-man collection team. Also, the engine was located inside the cab, just forward of the front axle.

The 1932 White Model 58SS with double-drop frame was specially designed for sanitation service. The 8-yard fully enclosed body in this photo was built by the White Company. Additional body types suitable for the chassis were available from other manufacturers. The truck shown here may be the pilot model in the huge New York City order for 774 similar units for delivery in 1932. Solid tires on new heavy-duty trucks were still specified by New York City up to 1933, due to their longer and cheaper service life.

A photo of the 1932 White Model 58SS with body raised provides a good view of the "double-drop" frame. The side-rails are evidently one-piece pressed steel, with the first drop under the cab and a kick-up over the rear axle. The frame drops behind the rear axle and is the location for the body tilting hinges. The wooden cab with peaked roof and sliding doors is very dated looking for 1932, but New York City was still using hundreds of trucks with a similar cab at the time.

A detail view of the Model 58SS White's rear axle area shows the massive Wood twin-piston telescoping hydraulic hoist. The base of the hoist is cradled between the chassis side-rails so it can make the necessary pivoting action as the body is raised. Also visible are the sizable twin brake drums acting on the propeller shaft, an important feature since there are no brakes on the front wheels. The rear wheels are cast steel and contain internal reduction gears as part of White's double reduction rear axle design.

A circa 1932 General Motors Truck, Model T55C, with 8-yard open body with side extensions. This view shows how mixed garbage and general household rubbish had to be stowed to make for an efficient refuse collection. A canvas cover is located on the cab's running board, ready for spreading over the open load. The cab door was most likely replaced during winter months, since easy access would be secondary to keeping the sanitation workers as warm as possible.

A Hendrickson Model 19S, circa 1933, with dump body having side extensions. This truck was designed for the speedy removal of moderate amounts of refuse from a suburban Chicago community. The cab is an up-dated design, which was introduced in the late 1920s, and blends well with the new stylized radiator grille and hood designs inspired by the latest automobile styles.

A 1935 Stewart 3 1/2- to 5-ton truck with Colecto Model S.H., 10- to 12-yard side-loader refuse body. By 1935 the basic Colecto side-loading body had been redesigned with the right, or loading side, now sloping into an overhang section. Also, by 1935 the Colecto body was being built in several rear-loading models having capacities up to 15 yards. The Stewart was a moderately priced truck, which was built in Buffalo, New York, up to the World War II era. This heavy Stewart represents styling first introduced in 1931.

The direct side view of the 1935 Stewart with Colecto body shows the loading bucket tucked into the body's overhang. The bucket was raised by cables in the body, and the body was raised by a traditional Wood combination hydraulic piston and cable hoist behind the cab. By 1936, Lynn, Massachusetts, was using 11 Colecto-bodied refuse trucks, with 6 being side-loaders and 5 that were rear loaders.

A White 1935 Model 702 chassis with a "Commercial" side-loading refuse body. This body was built by the Commercial Shearing and Stamping Company of Youngstown, Ohio, and was of moderate capacity. This firm built a line of 3-way hydraulic dump bodies designed for the use of road contractors. This style of White truck was built only in 1934 and 1935, being replaced with a streamlined 700-series truck-line designed by automotive stylist Alexis de Sakhnoffsky.

A 1935 AB Mack with Heil 10-yard duck-tail garbage body with side extensions. Most likely built for a private refuse contractor, the body has the canvas cover carrier jutting out over the cab's roof. Also, the truck has a radiator guard to fend off contacts with other vehicles or objects in tight spots, such as in alleyways and at dumpsites. The AB Mack was modernized in 1933 for those truck operators still wishing the basic economy of a 4-cylinder model.

A Model D90 1936 Sterling, double-reduction drive truck, with a custom-built 13-yard refuse body. This private sanitation contractor served a wholesale produce market in New York City known as the "Washington Market." Sterling trucks were built in Milwaukee, Wisconsin, in various capacities until the Depression era of the 1930s. From the late 1930s until the 1950s only heavy-duty trucks were offered. After a merger with the White Motor Company in 1951, Sterling-White trucks were built up to the end of 1953.

The Walter drop-frame refuse truck, Model FNQSA, was introduced in 1936. While its normal carrying capacity was 3 to 5 tons, the gross vehicle weight rating was 20,000 lbs. Since the Walter Motor Truck Company of Long Island, New York, specialized in multi-drive trucks, the new refuse chassis was a 4x4 and was designed for snow removal service too. Walter used a unique internal gear drive and locking differential to achieve what was called "4-Point Positive Drive," and "100% Traction."

A 1937 Model 21MB Mack, Jr. is seen here with a Heil arch-top Model SL-11 refuse body. With the flared sides and arch-top, the basic dump body's capacity was increased from 3 to 9 yards. The Mack, Jr. was built by the Reo Motor Car Company during the 1936 to 1938 period for the sale by Mack factory branches and distributors. The Mack, Jr. models provided a line of moderately priced light- and medium-duty trucks to compliment the more custom-built Mack medium- and heavy-duty Mack models.

This view of the 1937 Mack, Jr. shows how the curved doors on the side-loader slid over to the opposite side of the body. This style of sanitary refuse body was an answer to the objections raised to canvas covered open bodies. Depending upon the diligence of the collection crew, open bodied refuse trucks may not have been fully covered when filled, allowing refuse and papers to be scattered along city streets as the truck sped to the dump.

Cab-over-engine trucks had become very popular by 1937, and were now being considered for various sanitation services. The Model 818 White shown here was equipped with a 10-yard side-loading body. The heavy radiator grille guard and front frame extensions were add-ons for the unit to be used in winter snow removal service. This truck was part of the New York City order for 100 such units, and had a capacity rating of 3 to 6 tons.

This 1937 Walker heavy-duty battery-powered electric truck was equipped with a Heil 24-yard refuse body. With crowded New York City streets restricting the speed of traffic, the need for faster gasoline-powered trucks was overshadowed by the electric truck's great economy of operation. Many New York City delivery fleets had electric trucks at this time, and this electric refuse truck was most likely an experiment to see if a cost savings in such an application could be realized.

A Mack, Jr., Model 21MB, with Leach 6-yard "Garbage Getter" refuse body. The hopper on this rear loader was pulled up the back of the body by means of a chain and track system energized by the truck's power take-off. Also, a sliding door at the top of the body automatically opened to admit the refuse and closed to seal the body. The "Garbage Getter" was renamed the "Refuse Getter" soon after and continued in production into the 1950s.

Shown are two 1938 Model 89 Federal cab-over-engine trucks with Wood fully enclosed garbage bodies and twin-piston hydraulic hoists. The Federal Motor Truck Company of Detroit, Michigan, built moderately priced assembled trucks up to the 1950s. Grand Rapids, Michigan, received this pair of refuse collection trucks.

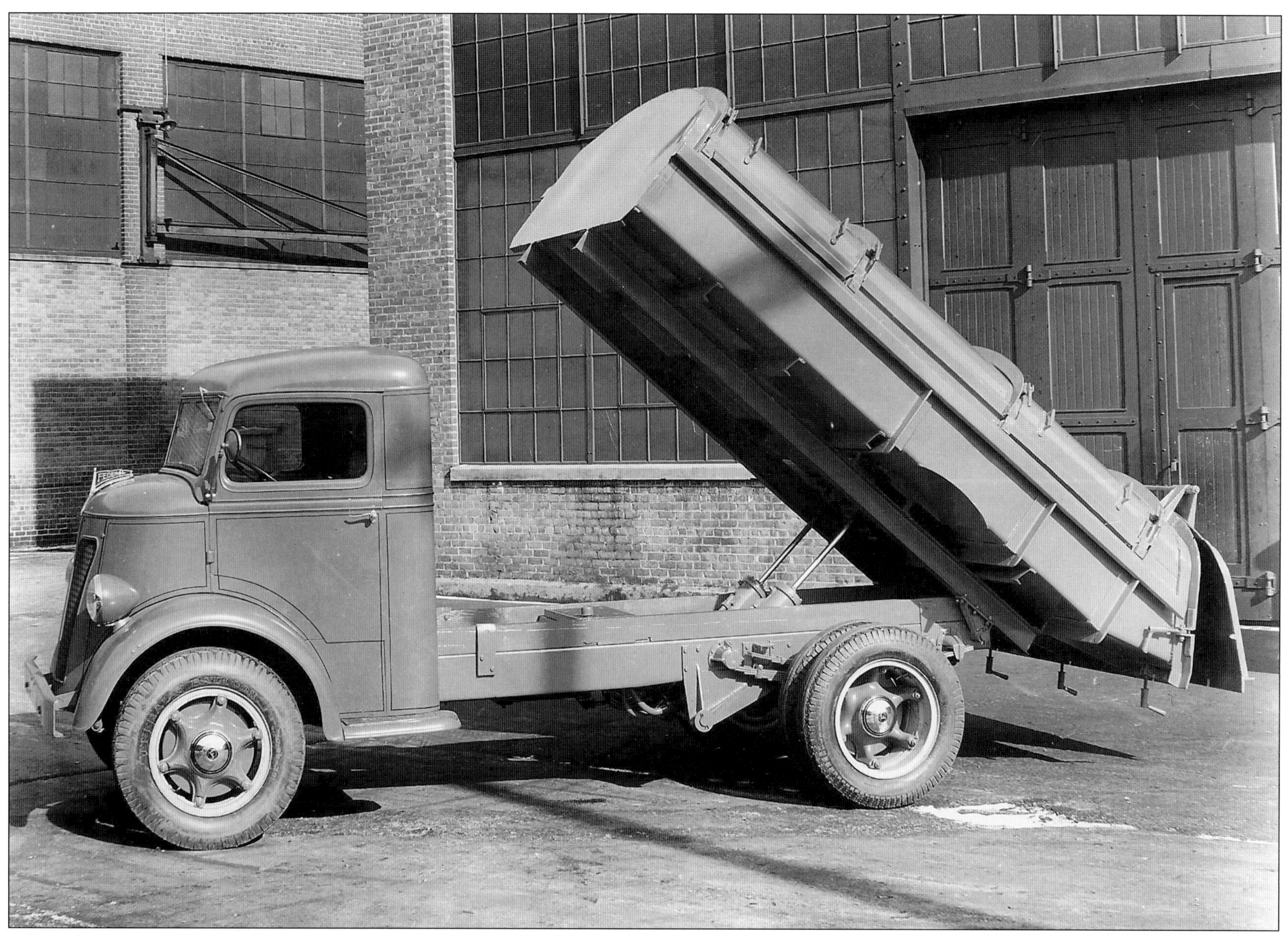

A side view of 1938 Federal cab-over-engine with Wood garbage body in raised position. This type of Wood body was built in various capacities from 4 to 6 yards. This view was most likely taken at the plant of the body builder, Gar Wood Industries, of Detroit, Michigan.

A rear view of the 1938 Federal with fully enclosed Wood body of about 6-yard capacity. Two sliding steel covers operated on steel rollers and extended across the body. The body is of the sump-type to retain liquids, and the tailgate was close fitting and considered watertight.

A special drop-frame 1938 GMC chassis with 21-yard Wood refuse body with hydraulically operated conveyor system. The body was raised by a Model T44 Wood dual cylinder underbody hoist. Also, the tailgate was raised hydraulically to allow discharge of the load. This vehicle has extra bumper brackets for the attachment of snow removal equipment, along with the older style cab used on cab-over-engine GMC trucks in the 1934 to 1936 period.

A direct side view of the 1938 GMC with Wood 21-yard refuse body. The tailgate is in the full-open position and shows the loading hopper touching the ground. The conveyor system ran up the tailgate and along the top of the body and operated continuously while refuse was being loaded into the hopper at the rear. It was powered by the truck's engine, and part of the drive system can be seen on top of the body at the front. New York City took delivery of 615 similar GMC refuse trucks in 1937, but the conveyor system was a new feature on the 1938 units.

A rear view of the 1938 GMC New York City garbage truck shows the refuse hopper. It is also a good view of the conveyor system, which speeded up the loading of these very large bodies. This system was not meant to compress the garbage, but more to distribute the material being stowed from one end of the body to the other as well as help confine offensive odors during the loading process. The 1938 New York City fleet order also included drop-frame Autocar trucks with conveyor-equipped bodies.

A Brockway 1938 M-series tractor with trailer-mounted special Wood Type W8 garbage and rubbish body. The 20-yard body was mounted on a special drop-frame Fruehauf semi-trailer, and elevated by a Wood underbody Model T33 dual cylinder telescopic hoist. The city of Rochester, New York, had a group of these units in their refuse collection fleet. The cab-forward Metropolitan series of Brockway trucks were specially designed for urban service, having a shorter turning radius and better driver visibility than conventional models.

A rear-loading Colecto body on a Model DR50 International, circa 1938. This Model 8 R.H. Colecto body had an 8-yard capacity. The hopper at the rear was raised by arms on each side of the body for discharge into the top of the body. A door at the top of the body automatically opened and closed during the bucket's discharge. A larger Model D Colecto rear-loader, 10- to 12-yards capacity, was also delivered to the city of Philadelphia at this time. The "D" model had an internal auger in the body, which compacted the load.

A heavy-duty 1939 Ward La France cab-over-engine chassis with side-loading Colecto Model S.H. body. This left-hand side view shows the cable device used to raise the loading bucket located on the right side of the body. Rutherford, New Jersey, used several Colecto-bodied trucks in its refuse collection fleet.

Photo taken in 1939 of FWD Model CU with 10-yard Heil refuse body. This body had a large rear hopper with right- and left-hand openings for both curb- and street-side loadings. The body came equipped with a chain-conveyor, which carried the refuse from the hopper to the interior of the body for continuous loading. The side door was needed for large rubbish items that could not be placed in the loading hopper. Also, while a 4x4 vehicle, there is no indication that the truck was intended for snowplow operation.

A 1939 Hendrickson Model A-275 tandem truck with Heil dump body having side extensions. The roof box on the cab was a convenient location for a tarpaulin to cover the load. This Chicago-area scavenger service needed a body with great volume, hence the extra side extensions, which were fully deployed in this view. The hood and grille styling on this Hendrickson was used on certain models during 1938 and 1939.

A 1939 International D-series truck with Galion arched-top refuse body. Semi-circular sliding doors on both sides of the body gave full access. This type of body was considered suitable for hauling various types of lightweight materials and bulky aggregate.

The 1939 International D-series with Galion arched-top body in the raised position. A Galion Model GH56 hydraulic hoist was most likely used with this style body. The Galion All Steel Body Company of Galion, Ohio, specialized in dump bodies for many years.

A 1940 V-8 Ford with 6 1/2-yard Gar Wood dump body and underbody hoist. This speedy medium-duty scavenger's truck served the South Side of Chicago, and carried a special extra tailgate for bulky rubbish items. Starting about 1938 the "Gar Wood" trade name was used in place of "Wood" on most of the company's products.

A 1940 Model D-500 International with a Gar Wood hydraulically operated "Load Packer" body. In 1938 Gar Wood introduced this rear-loading refuse body having a specially designed hopper-mounted tailgate, or pressure-plate, which compressed each load before it entered the body. This compression of each hopper loading allowed a body of moderate size to carry a much larger load. The inherent economy of this hydraulic compaction system resulted in the eventual popularity of the compactor-type refuse body.

A 1939 view of one of a group of ten Ward La France extra heavy-duty wreckers purchased by New York City. Gar Wood Industries supplied the heavy-duty boom and hydraulically operated power and control system. These recovery vehicles were essential after the first group of huge 21-yard garbage trucks were purchased in 1937. The municipal fleet of refuse and ash trucks numbered over 3,000 units at this time, and stranded trucks had to be recovered in an effective manner.

A 1945 Mack Model EHU with Gar Wood 16-yard Load-Packer body. Private sanitation services turned to the rear-loading compactor body after World War II when ordering new refuse trucks. This 1945 Model EHU Mack still has the chrome brightwork painted over, which was a wartime requirement on civilian trucks. Also, the EHU was the cab-over-engine version of the popular EH conventional model.

A rear view of EHU Mack with Gar Wood body shows larger postwar hopper construction. There are also steps on each side of the hopper for a two-man collection crew to stand between stops. Cab-over-engines, or "Traffic Types" as Mack marketed them, were considered advantageous for city traffic conditions due to their decrease in overall length and turning radius.

A postwar Brockway, Model 148L, with Heil Colecto-Pak 10-yard refuse body. This 1946 view shows the body in raised position for discharge. The two arms that raised the loading bucket to the top of the body also controlled the compression of the load. Cables attached to arms pulled on a pressure plate at the front of the body, which compressed each bucket load after its deposit in the body. Heil also built the Colecto body at this time, a rear-loader without the compaction feature.

A 1947 Mack Model EH with Gar Wood Load-Packer rear loader. Many private refuse haulers favored conventional engine-in-front trucks, due to their quicker engine accessibility and lower cab height for easier entry. The EH model was introduced in 1936 and built up to 1950, with a total production of 31,539 units. By the postwar period, the EH model had been reclassified through improvements from a medium-duty truck to one with a maximum 12-ton gross vehicle rating.

A Mack 1947 Model EQSW tandem chassis with Heil dump body with side extensions. During the early postwar period coal was still a major heat source for many buildings in New York City, and private carters removed many tons of ashes every day. Also, major public utilities burned coal in their power plants, and many coal consumers did not convert to oil until the late 1950s or even 1960s. The EQSW model was the largest of the E-series Macks and in the tandem version had a 20-ton maximum gross vehicle rating.

A Ward La France D-series truck, circa 1947, with Daybrook side-dump body of about 15-yard capacity. This hydraulically operated dump body was used to transport ashes from New York City garbage incinerators to various sanitary landfills. Before environmental concerns about wetlands, some marshy areas around New York City were filled in for park purposes or for other uses. The incinerators were gradually phased out as the dumpsites became filled, and growing concerns about air quality also forced a halt to many kinds of pollution emitting processes.

Model WB-18 White, delivered in 1948 with crew cab and Gar Wood 9-yard scow body. The cab certainly had room for at least four crew members, one sitting to the left of the driver. Several communities in the Cleveland, Ohio, area found the large open scow-type garbage body suitable for most of their refuse collection needs. This 1948 White was delivered to the city of Cleveland Heights.

This view of the 1948 White, Model WB-18, with body raised shows drop-frame chassis construction. A Gar Wood dual cylinder hydraulic hoist raised the body to the dump position. Two boxes under the body carried tools and a tarpaulin. Also, two semi-circular brackets were attached to the forward part of the body for carrying buckets needed at some collection sites.

A 1950 Dodge heavy-duty chassis with Leach Model 1-R, 15-yard Packmaster refuse body. In 1947 the Leach Company introduced a line of rear-loading compaction bodies, which were offered in 9-, 12-, 15- and 20-yard capacities by the mid-1950s. The heart of the Packmaster was a "bulldozer" blade that compressed each hopper load inside the body. Also, Dodge trucks were offered in maximum gross vehicle weights up to 28,000 lbs., more than adequate for the 15-yard Leach body weighing 8,500 lbs.

This 1950 Walter snow fighter, Model AEUL, served a dual purpose with the Department of Parks, New York City. With its large flotation-type tires, this Walter was able to negotiate sandy New York City beaches for rubbish collection purposes. A Marion Hydropaka side-loading hydraulic compaction body of about 15-yard capacity was mounted. During the winter a one-way speed plow and underbody center scraper were attached for clearing parkways around the city.

This St. Paul Pax-all, 12-yard compacting refuse body is mounted on a 1951 Model F-5 Ford truck chassis. The rear-mounted loading bucket was 6 feet wide, and dumped through a hatch on top of the body at the front. A "packer panel" at the front of the body automatically pushed each new bucket load toward the back of the body. The body tilted during this combined dump and compaction process. The St. Paul Company was owned by Gar Wood, and built Pax-all refuse bodies in 9- and 12-yard capacities.

A 1951 Mack Model A40H with Gar Wood Load-Packer refuse body for use in contract rubbish and garbage collection service. The A40H model was introduced in 1950 and replaced the EH Mack with the same 12-ton gross vehicle rating. The Mack A-series trucks were interim models, which used sheet metal styling patterned after the larger L-series.

The 1951 Mack Model A40H with Gar Wood body raised to dump position by a dual cylinder hoist. This truck and body without the payload weighed 8 tons, leaving a 4-ton allowance for the basic load without overloading. The Mack A-series was replaced by a new B-series in 1953.

A heavy-duty 1951 Model U-50 Autocar with 15-yard Gar Wood refuse body. This unit was delivered to the Philadelphia suburban Township of Upper Darby, Pennsylvania. Also, the cab-over-engine Autocar had a gross vehicle weight of 25,000 lbs. and a wheelbase of only 124 inches. It is shown dumping a rubbish collection at a local landfill site.

A circa-1950 FWD HR-series 4x4 truck with Gar Wood Load-Packer refuse body. The HR was considered a medium- to heavy-duty model with a nominal gross vehicle weight rating of 20,000 lbs. However, this rating was subject to variation with the type of application. This unit served the lower Hudson Valley community of Hastings-on-Hudson, and may have also been used in snowplow service.

A 1953 FWD Model HRY 4x4 truck with Gar Wood Load-Packer refuse body. The operator of the major New York/ New Jersey bridges and tunnels, the Port of New York Authority, owned this vehicle. It had the bumper-mounted frame hitch for the mounting and control of a snowplow. Special cab-mounted warning lights were needed to alert traffic during the slow movements of the truck at night and in stormy weather.

A 1953 White 3000-series truck with a Roto-Pac 19-yard body with roof rack for large rubbish items. The Roto-Pac refuse body was called an "escalator-compactor" type. A low-hung hopper at the body's rear fed refuse to a continuously moving escalator, which in turn moved the material to an opening near the top of the body. Located inside the opening was a rotary compaction plate, which continuously compressed and forced the mangled refuse deeper into the body. The minimum gross vehicle weight given for this refuse body with chassis was 24,000 lbs.

A 16-yard Roto-Pac refuse body on a 1953 Ford 900-series "Big Job." This Roto-Pac body was equipped with twin telescopic two-sleeve hydraulic hoists. Each of the underbody hoists was mounted outboard of the chassis siderails, just forward of the rear tires. Roto-Pac bodies were built by the City Tank Corporation in New York City. During the early 1950s Roto-Pac bodies were built in 16- and 19-yard capacities. The minimum gross vehicle weight for this body and chassis was considered to be about 22,000 lbs.

Another Walter "Snow Fighter" refuse truck was acquired for the New York City beach cleaning in 1953. Like the first unit, this one also sported large flotation-type tires, but had a 16-yard Gar Wood Load-Packer body. This unit also cleared New York City area parkways during winter months. While Walter was well known for its line of snow fighting and highway maintenance trucks, starting in 1927 many types of single and dual purpose refuse trucks were built by the company.

A fleet of 5 Diamond T Model 422 refuse trucks were delivered to the city of Warwick, Rhode Island, in 1953. The Diamond T Model 422 was rated as a 3- to 5-ton load carrier, but could be modified for severe applications. Diamond T also produced a cab-forward tilt-cab model that could be modified for refuse truck service. The bodies on the Warwick refuse trucks are Leach1-R models of about 15-yard capacity.

A 1954 Autocar C-series with 16-yard Roto-Pac body. This New York City scene shows a typical garbage pickup in the Borough of Manhattan, New York City. The sanitation workers are unloading burlap bags of mixed garbage and rubbish into the Roto-Pac's hopper. New York purchased 200 Autocar refuse trucks during 1947 and 1948, and another group in 1954. Both Load-Packer and Roto-Pac bodies were acquired with the Autocars.

A Mack Model N-42 low cab-forward, circa 1960, with Dempster Model CA30-24DB Dumpmaster 24-yard body. The N-series Mack was introduced in 1958, using a cab developed for the Ford C-series truck line by the Budd Company. The Dempster Dumpmaster line of front-loaders included 18- and 30-yard models at this time. These bodies incorporated a telescopic cylinder and packer plate for full load compaction. The Dempster front-loaders were designed to pick up Dempster refuse containers left at pick-up sites.

The Dempster "Dinopacker" Type SS-428 compaction body of 1960 was an early type of roll-off container suitable for refuse collections. The roll-off container concept seems to be a development of the late 1950s, as an extension of the smaller portable containers used for many years for the convenient handling of one and two yards of material. The roll-off container was actually a portable truck body, which saved the time and expense of a refuse truck making many stops at the same location during a week or more time period.

The 1961 Brockway, Model 148LL (shown here) has been equipped with a Leach Model 2-R refuse body. This heavy-duty Brockway tandem chassis had a stated gross vehicle weight of 40,000 lbs., which was needed with the large Model 2-R rear-loading refuse truck. The Leach 2-R combined a large body and rugged components to handle many kinds of rubbish, including demolition material. The body's compaction pressure was also increased, and it had a huge hopper that could take rear-loading containers up to 10 yards.

This 1962 Brockway Model 148L came equipped with a Model 50K Delux cab and had a gross vehicle weight rating of 28,000 to 30,000 lbs. The Heil Colectomatic, Mark II refuse body had several critical improvements over its predecessor model. It used an ejector panel for unloading the body, which avoided the instability problems of the trucks having body hoists when they were raised for unloading at landfills. The addition of the ejector panel along with the regular packer panel resulted in a "double squeeze." This was called a duo-press, which resulted in an estimated 15% to 20% greater compaction than for a packer mechanism acting alone.

A custom-built Hendrickson tilt-cab chassis, circa 1962, with Leach Packmaster 20-yard refuse body. The walk-in cab had sliding doors and was set forward of the front axle, allowing a lower height and shortening the wheelbase for an improved turning radius. The Hendrickson chassis was powered by a GMC Model 4-53 diesel engine, and the gross vehicle weight was given as 49,000 lbs.

A White Model 4400D with Leach Packmaster body, circa 1963, owned by a Florida-based private sanitation service. The White heavy-duty 4000-series was introduced in 1956, and lighter-duty models in the 2000-series in 1958. Between 1953 and 1958 The White Motor Corporation had taken over the Autocar, Reo, and Diamond T lines of trucks, each of which offered heavy-duty models suitable for refuse collection services.

This Oshkosh Model W-616 with Leach Packmaster 20-yard body delivered to Yonkers, New York, in 1963. This city, just north of New York City, is known for its many hills, and this 4x4 Oshkosh was fitted for dual service, as it also plowed streets in the winter. The truck's builder, Oshkosh Motor Truck, Inc., of Oshkosh, Wisconsin, specialized in multi-drive commercial vehicles. This Model W-616 had a gross vehicle weight of 36,000 lbs., with the basic chassis portion with standard equipment being 11,180 lbs.

This action view shows a Dempster "Dumpmaster" fitted to a 1964 Ford Model C-950. A large container, designed for industrial purposes, has been raised to dump position by the lifting forks after the lifting arms have rotated backward. This particular Dumpmaster body does not have the compaction mechanism used on regular refuse bodies.

After dumping the container and returning it to its ground location, the lifting forks and arms have been placed into "travel-position." The Dempster Brothers Company of Knoxville, Tennessee, had developed a system of portable material containers for the contracting industry during the 1930s. The Dempster Brothers Company was purchased by the Carrier Corporation in 1970 and the name changed to Dempster Dumpster Systems in 1975.

The chassis for the Wesco-Jet front loader was developed by Reo in 1966. The lifting arms of the front loading device operated between the two cab sections for maximum safety. Also, the extreme forward overhang of the cab shifted enough weight to the front to achieve a total payload of up to 12 1/2 tons without the use of tandem rear axles. When the Reo truck line became the Diamond-Reo in 1967, this chassis became the Model CF0-5542 and was supplied to the West Coast builder of the refuse body until about 1968.

A 1973 Brockway, Model 358L, with a cylindrical 23-yard Truxmore Pakker refuse body. The body was a side-loader, which could be filled from both sides at the same time when conditions allowed. The Truxmore Pakker was a compactor-type refuse body, which was also built in 18- and 27-yard sizes. This Brockway was owned by the city of Cortland, New York, where it also helped to clear 50 miles of streets during the winter months.

An H-series Hendrickson tandem, circa 1974, with Leach Packmaster Model 2R refuse body. The Hendrickson Manufacturing Company continued to make custom trucks into the 1980s through its Mobile Equipment Division. During the 1980s this operation concentrated more on custom chassis lines for special equipment, such as fire apparatus and mobile cranes.

A heavy-duty 1975 Mack MB-series tandem with huge "E-Z Pack" 40-yard refuse body. The front-loader combined a 33-yard main body with compaction equipment and a 7-yard tailgate bubble for a 40-yard total. This contract hauler served private commercial customers in some urban areas of Connecticut. The cab-forward MB-series was introduced for city delivery work in 1963 and phased out in 1978.

A Mack 1976 Model MB400 tandem with 40-yard Amrep roll-off container. This high-sided, open top container was designed basically as a general rubbish and demolition material carrier. Access to such containers was usually through a vertically hinged tailgate and open top. The truck was equipped with tilting rails and a power loading/unloading device, which pulled the container onto the chassis or let it roll down at a drop-site. This unit was used by a private carter in Southern California.

A late-1976 Brockway Model 550LL with Gar Wood, Model 900, 25-yard rear-loader refuse body. This tandem model was purchased for residential curb collections in the Galesburg, Illinois, area and was fitted with an automatic transmission. The cab-forward Brockway "Huskiteer" line was introduced in 1971, and was designed primarily for urban delivery service. It was built up to 1977, the year when the Brockway plant in Cortland, New York, ceased operation.

A special 1977 Walter front-drive tandem chassis with Pak-Mor rear-loader refuse body. The truck also had right-hand control and the cab was equipped with sliding doors for both driver and helper. Positioning of the cab forward of the front (drive) axle shifted more weight to this axle, helping to produce more traction. In 1957 the Walter truck operation left its plant on Long Island, New York, and moved upstate to a larger facility in the Albany area. At the new plant in Voorheesville, Walter built this and several other interesting pieces of refuse equipment.

A Maxon Eagle SL side-loader, 29-yard compaction refuse vehicle delivered to Mt. Vernon, Illinois, in 1979. Maxon Industries engineered a chassis, cab and body dedicated to refuse collection. The Maxon "Eagle" SL "Side-Pak" was the initial vehicle, which series were planned to include units of 21, 25 and 29 yards on a single rear axle, and 33 and 37 yards on tandem rear suspensions. The Maxon SL chassis had a drop-frame and Hydraulically powered tilt-cab.

A 1980 American LaFrance Model CTC refuse chassis with 25-yard E-Z Pack rear-loading body. The Model CTC was a basic cab-forward design, which was built in both single and tandem rear axle configurations. The LECTC was a low-entry model, which had the cab placed fully ahead of the front axle. This achieved easier entry and a better weight distribution. The LECTC was available in a drop-frame version for side-loading refuse bodies.

A White cab-forward, 1981 Xpeditor 2, Model RX2-64 with Amrep front loader refuse body. Amrep Incorporated, of Ontario, California, supplied bodies and refuse collection equipment to many West Coast communities. The White Road Xpeditor 2 was introduced in 1977 as a new cab design for the prior Road Xpeditor, which in turn had used a modified White Compact-series cab and chassis when it was announced in 1974.

A 1981 White Xpeditor 2 chassis which has been modified with a drop-frame by Amrep to take a side-loader body. At this time the White Xpeditor 2 chassis was not available with drop-frame construction, and any such design changes had to be done in the field. This view of the modified chassis with body slightly raised shows the rear of the cab with all the body controls in place.

An American LaFrance Model CTC, circa 1982, with Lodal front-loading body of 36- to 46-yards capacity. It also has been equipped for the Lodal "Load-A-Matic" container system. The hydraulic loading arms are seen here in their normal travel position. During the early 1980s American LaFrance also built the "Gold Star," Model LE, low entry chassis with optional drop-frame construction for side-loading bodies.

A Mack MR600 series tandem, circa 1982, with tag-axle, making a tri-axle truck. The extra axle was most likely needed to meet local state axle-weight laws, due to the high capacity of the huge front-loader body. This unit is posed with its Load-A-Matic loading arms ready to lift a wheeled container. Lodal, Inc., of Kingsford, Michigan, built various types of heavy refuse collection equipment at this time.

A Peterbilt 300-series chassis with "E-Z Pack" roll-off container rails in raised position. Twin single-sleeve hydraulic hoists have positioned the rails in a 45-degree angle. The rail ends have a roller to avoid snagging the device when backing up to a container located near or on irregular ground. There is also a heavy cable for hooking to the container during the loading operation. Peterbilt offered various trucks for use in refuse collection work, including the 320 and 330 models.

A Crane Carrier low entry chassis, circa 1982, with E-Z Pack 29-yard side-loader body. This CCC Centurian chassis has a drop-frame design, with the cab being set up with dual controls for either left- or right-hand operation. The Crane Carrier Corporation, of Tulsa, Oklahoma, introduced the Centurian "Low Profile Carrier" in 1974, with various design improvements since then. Crane Carrier has also built complete refuse collection vehicles, including side-loaders and roll-off container carriers.

A special drop-frame 1982 Mack Model MR600 side-loader chassis with E-Z Pack 37-yard body. The MR side-loader chassis also incorporated a galvanized cab with dual left- and right-hand controls. The right-hand controls also involved a stand-up position for a more efficient one-man operation. The Peabody Galion Company of Galion, Ohio, built E-Z Pack bodies in a variety of capacities at this time. McClain Industries took over the Galion operation and continued to produce the E-Z Pack line of refuse equipment.

A 1986 FWD Model A64-3221 low-cab forward chassis with 25-yard Leach body. The FWD Corporation introduced its "America" line of specially designed vehicles for refuse collection services in 1986. Both single rear axle and tandem rear axle models were offered, with gross vehicle ratings from 43,000 to 58,000 lbs. The Leach-bodied FWD shown here was delivered to the city of Green Bay, Wisconsin.

An Ottawa 1988 “Rogue” series refuse chassis with large Dempster rear-loader compactor body. Ottawa offered the Rogue chassis in four models: straight-frame units with single or tandem rear axles, and drop-frame units with single or tandem rear axles. The Ottawa Truck Corporation introduced its Rogue chassis line about 1987, but sold it to the Oshkosh Truck Corporation in 1992.

A Mack MS, Mid-Liner series, with Dempster rear-loader body, circa 1988. The Mack Mid-Liner was announced in 1979, and was based on a truck model built by the Renault operation in France. As a mid-range diesel designed for city delivery work, it found some utility in refuse collection service.

The 1991 FWD Model MF42-3321 had low-entry cab set forward of the front axle. This design distributed more of the body and payload weight to the front axle. With this type of weight distribution it was possible to use larger capacity bodies without the use of more expensive tandem suspensions. Trucks with this type of chassis design were also considered to be more maneuverable in city streets. The FWD Corporation's special refuse chassis program was started in 1986 and phased out during the 1990s.

A Mack Model MR tandem with Heil DuraPack Half/Pac front-loader. The basic body capacity of 28 yards is augmented by a 12-yard hopper, bringing the total to 40 yards. The unit shown here has a dumping mechanism, but another version is called a "full-ejection" model, which empties the body without tilting it.

A Volvo Xpeditor chassis with Heil 28-yard DuraPack compacting body with "Rapid Rail" automated side-loader. The automated side-loader is becoming more popular with the constant need to keep the lid on refuse collection costs. It can also provide for a more efficient sanitation system with the collection authority providing standardized containers to the customers.

A Volvo Xpeditor tractor pulling a trailerized Heil "STARR" automated collection unit. The trailer's body has a 33-yard capacity, and is filled by a Heil "Rapid Rail" automated side-loader located behind the cab of the tractor. The articulation of the tractor and trailer provides for a tight turning radius, and the ability to make collections in difficult places. When filled, the trailer can be detached, and, through the use of dolly wheels, hooked to another trailer for transport to the landfill.

A Sterling Acterra chassis with Heil "Retriever" side-loader and 10-yard body. Conceived for less accessible gated communities, parks, and recreational areas, the Retriever was designed to dump its payload into a full-size rear loader, and then continue on its own route. The Sterling Truck Corporation is a subsidiary of the Freightliner Corporation, and introduced the Acterra series in late 1999.

MORE TITLES FROM ICONOGRAFIX:

AMERICAN CULTURE
AMERICAN SERVICE STATIONS 1935-1943 PHOTO ARCHIVE ... ISBN 1-882256-27-1
COCA-COLA: A HISTORY IN PHOTOGRAPHS 1930-1969 ... ISBN 1-882256-46-8
COCA-COLA: ITS VEHICLES IN PHOTOGRAPHS 1930-1969 ... ISBN 1-882256-47-6
PHILLIPS 66 1945-1954 PHOTO ARCHIVE ... ISBN 1-882256-42-5

AUTOMOTIVE
CADILLAC 1948-1964 PHOTO ALBUM ... ISBN 1-882256-83-2
CAMARO 1967-2000 PHOTO ARCHIVE ... ISBN 1-58388-032-1
CLASSIC AMERICAN LIMOUSINES 1955-2000 PHOTO ARCHIVE ... ISBN 1-58388-041-0
CORVETTE THE EXOTIC EXPERIMENTAL CARS, LUDVIGSEN LIBRARY SERIES ... ISBN 1-58388-017-8
CORVETTE PROTOTYPES & SHOW CARS PHOTO ALBUM ... ISBN 1-882256-77-8
EARLY FORD V-8S 1932-1942 PHOTO ALBUM ... ISBN 1-882256-97-2
IMPERIAL 1955-1963 PHOTO ARCHIVE ... ISBN 1-882256-22-0
IMPERIAL 1964-1968 PHOTO ARCHIVE ... ISBN 1-882256-23-9
LINCOLN MOTOR CARS 1920-1942 PHOTO ARCHIVE ... ISBN 1-882256-57-3
LINCOLN MOTOR CARS 1946-1960 PHOTO ARCHIVE ... ISBN 1-882256-58-1
PACKARD MOTOR CARS 1935-1942 PHOTO ARCHIVE ... ISBN 1-882256-44-1
PACKARD MOTOR CARS 1946-1958 PHOTO ARCHIVE ... ISBN 1-882256-45-X
PONTIAC DREAM CARS, SHOW CARS & PROTOTYPES 1928-1998 PHOTO ALBUM ... ISBN 1-882256-93-X
PONTIAC FIREBIRD TRANS-AM 1969-1999 PHOTO ALBUM ... ISBN 1-882256-95-6
PONTIAC FIREBIRD 1967-2000 PHOTO HISTORY ... ISBN 1-58388-028-3
STUDEBAKER 1933-1942 PHOTO ARCHIVE ... ISBN 1-882256-24-7
ULTIMATE CORVETTE TRIVIA CHALLENGE ... ISBN 1-58388-035-6

BUSES
BUSES OF MOTOR COACH INDUSTRIES 1932-2000 PHOTO ARCHIVE ... ISBN 1-58388-039-9
THE GENERAL MOTORS NEW LOOK BUS PHOTO ARCHIVE ... ISBN 1-58388-007-0
GREYHOUND BUSES 1914-2000 PHOTO ARCHIVE ... ISBN 1-58388-027-5
MACK® BUSES 1900-1960 PHOTO ARCHIVE* ... ISBN 1-58388-020-8
TRAILWAYS BUSES 1936-2001 PHOTO ARCHIVE ... ISBN 1-58388-029-1

EMERGENCY VEHICLES
AMERICAN LAFRANCE 700 SERIES 1945-1952 PHOTO ARCHIVE ... ISBN 1-882256-90-5
AMERICAN LAFRANCE 700 SERIES 1945-1952 PHOTO ARCHIVE VOLUME 2 ... ISBN 1-58388-025-9
AMERICAN LAFRANCE 700 & 800 SERIES 1953-1958 PHOTO ARCHIVE ... ISBN 1-882256-91-3
AMERICAN LAFRANCE 900 SERIES 1958-1964 PHOTO ARCHIVE ... ISBN 1-58388-002-X
CLASSIC AMERICAN AMBULANCES 1900-1979 PHOTO ARCHIVE ... ISBN 1-882256-94-8
CLASSIC AMERICAN FUNERAL VEHICLES 1900-1980 PHOTO ARCHIVE ... ISBN 1-58388-016-X
CLASSIC SEAGRAVE 1935-1951 PHOTO ARCHIVE ... ISBN 1-58388-034-8
FIRE CHIEF CARS 1900-1997 PHOTO ALBUM ... ISBN 1-882256-87-5
HEAVY RESCUE TRUCKS 1931-2000 PHOTO GALLERY ... ISBN 1-58388-045-3
LOS ANGELES CITY FIRE APPARATUS 1953 - 1999 PHOTO ARCHIVE ... ISBN 1-58388-012-7
MACK MODEL C FIRE TRUCKS 1957-1967 PHOTO ARCHIVE* ... ISBN 1-58388-014-3
MACK MODEL CF FIRE TRUCKS 1967-1981 PHOTO ARCHIVE* ... ISBN 1-882256-63-8
MACK MODEL L FIRE TRUCKS 1940-1954 PHOTO ARCHIVE* ... ISBN 1-882256-86-7
NAVY & MARINE CORPS FIRE APPARATUS 1836 -2000 PHOTO GALLERY ... ISBN 1-58388-031-3
PIERCE ARROW FIRE APPARATUS 1979-1998 PHOTO ARCHIVE ... ISBN 1-58388-023-2
POLICE CARS: RESTORING, COLLECTING & SHOWING AMERICA'S FINEST SEDANS ISBN 1-58388-046-1
SEAGRAVE 70TH ANNIVERSARY SERIES PHOTO ARCHIVE ... ISBN 1-58388-001-1
VOLUNTEER & RURAL FIRE APPARATUS PHOTO GALLERY ... ISBN 1-58388-005-4
WARD LAFRANCE FIRE TRUCKS 1918-1978 PHOTO ARCHIVE ... ISBN 1-58388-013-5
YOUNG FIRE EQUIPMENT 1932-1991 PHOTO ARCHIVE ... ISBN 1-58388-015-1

RACING
GT40 PHOTO ARCHIVE ... ISBN 1-882256-64-6
INDY CARS OF THE 1950s, LUDVIGSEN LIBRARY SERIES ... ISBN 1-58388-018-6
INDIANAPOLIS RACING CARS OF FRANK KURTIS 1941-1963 PHOTO ARCHIVE ... ISBN 1-58388-026-7
JUAN MANUEL FANGIO WORLD CHAMPION DRIVER SERIES PHOTO ALBUM ... ISBN 1-58388-008-9
LE MANS 1950: THE BRIGGS CUNNINGHAM CAMPAIGN PHOTO ARCHIVE ... ISBN 1-882256-21-2
MARIO ANDRETTI WORLD CHAMPION DRIVER SERIES PHOTO ALBUM ... ISBN 1-58388-009-7
NOVI V-8 INDY CARS 1941-1965 KARL LUDVIGSEN LIBRARY SERIES ... ISBN 1-58388-037-2
SEBRING 12-HOUR RACE 1970 PHOTO ARCHIVE ... ISBN 1-882256-20-4
VANDERBILT CUP RACE 1936 & 1937 PHOTO ARCHIVE ... ISBN 1-882256-66-2

RAILWAYS
CHICAGO, ST. PAUL, MINNEAPOLIS & OMAHA RAILWAY 1880-1940 PHOTO ARCHIVE ISBN 1-882256-67-0
CHICAGO & NORTH WESTERN RAILWAY 1975-1995 PHOTO ARCHIVE ... ISBN 1-882256-76-X
GREAT NORTHERN RAILWAY 1945-1970 PHOTO ARCHIVE ... ISBN 1-882256-56-5
GREAT NORTHERN RAILWAY 1945-1970 VOL 2 PHOTO ARCHIVE ... ISBN 1-882256-79-4
MILWAUKEE ROAD 1850-1960 PHOTO ARCHIVE ... ISBN 1-882256-61-1
MILWAUKEE ROAD DEPOTS 1856-1954 PHOTO ARCHIVE ... ISBN 1-58388-040-2
SHOW TRAINS OF THE 20TH CENTURY ... ISBN 1-58388-030-5
SOO LINE 1975-1992 PHOTO ARCHIVE ... ISBN 1-882256-68-9
TRAINS OF THE TWIN PORTS, DULUTH-SUPERIOR IN THE 1950s PHOTO ARCHIVE ... ISBN 1-58388-003-8
TRAINS OF THE CIRCUS 1872-1956 PHOTO ARCHIVE ... ISBN 1-58388-024-0
TRAINS OF THE UPPER MIDWEST: STEAM & DIESEL IN THE 1950S & 1960S ... ISBN 1-58388-036-4
WISCONSIN CENTRAL LIMITED 1987-1996 PHOTO ARCHIVE ... ISBN 1-882256-75-1
WISCONSIN CENTRAL RAILWAY 1871-1909 PHOTO ARCHIVE ... ISBN 1-882256-78-6

TRUCKS
BEVERAGE TRUCKS 1910-1975 PHOTO ARCHIVE ... ISBN 1-882256-60-3
BROCKWAY TRUCKS 1948-1961 PHOTO ARCHIVE* ... ISBN 1-882256-55-7
CHEVROLET EL CAMINO PHOTO HISTORY INCL GMC SPRINT & CABALLERO ... ISBN 1-58388-044-5
DODGE PICKUPS 1939-1978 PHOTO ALBUM ... ISBN 1-882256-82-4
DODGE POWER WAGON PHOTO HISTORY ... ISBN 1-58388-019-4
DODGE TRUCKS 1929-1947 PHOTO ARCHIVE ... ISBN 1-882256-36-0
DODGE TRUCKS 1948-1960 PHOTO ARCHIVE ... ISBN 1-882256-37-9
FORD HEAVY DUTY TRUCKS 1948-1998 PHOTO HISTORY ... ISBN 1-58388-043-7
JEEP 1941-2000 PHOTO ARCHIVE ... ISBN 1-58388-021-6
JEEP PROTOTYPES & CONCEPT VEHICLES PHOTO ARCHIVE ... ISBN 1-58388-033-X
LOGGING TRUCKS 1915-1970 PHOTO ARCHIVE ... ISBN 1-882256-59-X
MACK MODEL AB PHOTO ARCHIVE* ... ISBN 1-882256-18-2
MACK AP SUPER-DUTY TRUCKS 1926-1938 PHOTO ARCHIVE* ... ISBN 1-882256-54-9
MACK MODEL B 1953-1966 VOL 1 PHOTO ARCHIVE* ... ISBN 1-882256-19-0
MACK MODEL B 1953-1966 VOL 2 PHOTO ARCHIVE* ... ISBN 1-882256-34-4
MACK EB-EC-ED-EE-EF-EG-DE 1936-1951 PHOTO ARCHIVE* ... ISBN 1-882256-29-8
MACK EH-EJ-EM-EQ-ER-ES 1936-1950 PHOTO ARCHIVE* ... ISBN 1-882256-39-5
MACK FC-FCSW-NW 1936-1947 PHOTO ARCHIVE* ... ISBN 1-882256-28-X
MACK FG-FH-FJ-FK-FN-FP-FT-FW 1937-1950 PHOTO ARCHIVE* ... ISBN 1-882256-35-2
MACK LF-LH-LJ-LM-LT 1940-1956 PHOTO ARCHIVE* ... ISBN 1-882256-38-7
MACK TRUCKS PHOTO GALLERY* ... ISBN 1-882256-88-3
NEW CAR CARRIERS 1910-1998 PHOTO ALBUM ... ISBN 1-882256-98-0
PLYMOUTH COMMERCIAL VEHICLES PHOTO ARCHIVE ... ISBN 1-58388-004-6
REFUSE TRUCKS PHOTO ARCHIVE ... ISBN 1-58388-042-9
STUDEBAKER TRUCKS 1927-1940 PHOTO ARCHIVE ... ISBN 1-882256-40-9
STUDEBAKER TRUCKS 1941-1964 PHOTO ARCHIVE ... ISBN 1-882256-41-7
WHITE TRUCKS 1900-1937 PHOTO ARCHIVE ... ISBN 1-882256-80-8

TRACTORS & CONSTRUCTION EQUIPMENT
CASE TRACTORS 1912-1959 PHOTO ARCHIVE ... ISBN 1-882256-32-8
CATERPILLAR PHOTO GALLERY ... ISBN 1-882256-70-0
CATERPILLAR POCKET GUIDE THE TRACK-TYPE TRACTORS 1925-1957 ... ISBN 1-58388-022-4
CATERPILLAR D-2 & R-2 PHOTO ARCHIVE ... ISBN 1-882256-99-9
CATERPILLAR D-8 1933-1974 INCLUDING DIESEL 75 & RD-8 PHOTO ARCHIVE ... ISBN 1-882256-96-4
CATERPILLAR MILITARY TRACTORS VOLUME 1 PHOTO ARCHIVE ... ISBN 1-882256-16-6
CATERPILLAR MILITARY TRACTORS VOLUME 2 PHOTO ARCHIVE ... ISBN 1-882256-17-4
CATERPILLAR SIXTY PHOTO ARCHIVE ... ISBN 1-882256-05-0
CATERPILLAR TEN INCLUDING 7C FIFTEEN & HIGH FIFTEEN PHOTO ARCHIVE ... ISBN 1-58388-011-9
CATERPILLAR THIRTY 2ND ED. INC. BEST THIRTY, 6G THIRTY & R-4 PHOTO ARCHIVE ISBN 1-58388-006-2
CLETRAC AND OLIVER CRAWLERS PHOTO ARCHIVE ... ISBN 1-882256-43-3
CLASSIC AMERICAN STEAMROLLERS 1871-1935 PHOTO ARCHIVE ... ISBN 1-58388-038-0
FARMALL CUB PHOTO ARCHIVE ... ISBN 1-882256-71-9
FARMALL F- SERIES PHOTO ARCHIVE ... ISBN 1-882256-02-6
FARMALL MODEL H PHOTO ARCHIVE ... ISBN 1-882256-03-4
FARMALL MODEL M PHOTO ARCHIVE ... ISBN 1-882256-15-8
FARMALL REGULAR PHOTO ARCHIVE ... ISBN 1-882256-14-X
FARMALL SUPER SERIES PHOTO ARCHIVE ... ISBN 1-882256-49-2
FORDSON 1917-1928 PHOTO ARCHIVE ... ISBN 1-882256-33-6
HART-PARR PHOTO ARCHIVE ... ISBN 1-882256-08-5
HOLT TRACTORS PHOTO ARCHIVE ... ISBN 1-882256-10-7
INTERNATIONAL TRACTRACTOR PHOTO ARCHIVE ... ISBN 1-882256-48-4
INTERNATIONAL TD CRAWLERS 1933-1962 PHOTO ARCHIVE ... ISBN 1-882256-72-7
JOHN DEERE MODEL A PHOTO ARCHIVE ... ISBN 1-882256-12-3
JOHN DEERE MODEL B PHOTO ARCHIVE ... ISBN 1-882256-01-8
JOHN DEERE MODEL D PHOTO ARCHIVE ... ISBN 1-882256-00-X
JOHN DEERE 30 SERIES PHOTO ARCHIVE ... ISBN 1-882256-13-1
MINNEAPOLIS-MOLINE U-SERIES PHOTO ARCHIVE ... ISBN 1-882256-07-7
OLIVER TRACTORS PHOTO ARCHIVE ... ISBN 1-882256-09-3
RUSSELL GRADERS PHOTO ARCHIVE ... ISBN 1-882256-11-5
TWIN CITY TRACTOR PHOTO ARCHIVE ... ISBN 1-882256-06-9

All Iconografix books are available from direct mail specialty book dealers and bookstores worldwide, or can be ordered from the publisher. For book trade and distribution information or to add your name to our mailing list and receive a **FREE CATALOG** contact:

Iconografix, PO Box 446, Hudson, Wisconsin, 54016   Telephone: (715) 381-9755, (800) 289-3504 (USA), Fax: (715) 381-9756

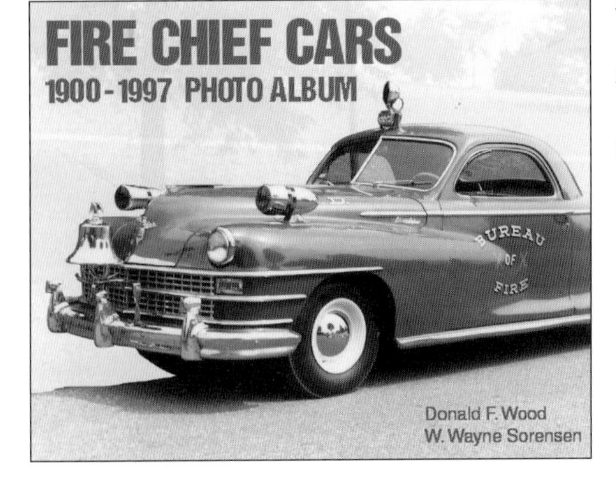

MORE GREAT BOOKS FROM ICONOGRAFIX

NEW CAR CARRIERS 1910-1998 PHOTO ALBUM
ISBN 1-882256-98-0

DODGE POWER WAGONS 1940-1980 PHOTO ARCHIVE
ISBN 1-882256-89-1

FIRE CHIEF CARS 1900-1997 PHOTO ALBUM
ISBN 1-882256-87-5

MACK MODEL AB PHOTO ARCHIVE*
ISBN 1-882256-18-2

MACK MODEL B 1953-1966 VOL 1 PHOTO ARCHIVE*
ISBN 1-882256-19-0

LOGGING TRUCKS 1915-1970 PHOTO ARCHIVE
ISBN 1-882256-59-X

BEVERAGE TRUCKS 1910-1975 PHOTO ARCHIVE
ISBN 1-882256-60-3

ICONOGRAFIX, INC.
P.O. BOX 446, DEPT BK,
HUDSON, WI 54016
FOR A FREE CATALOG CALL:
1-800-289-3504

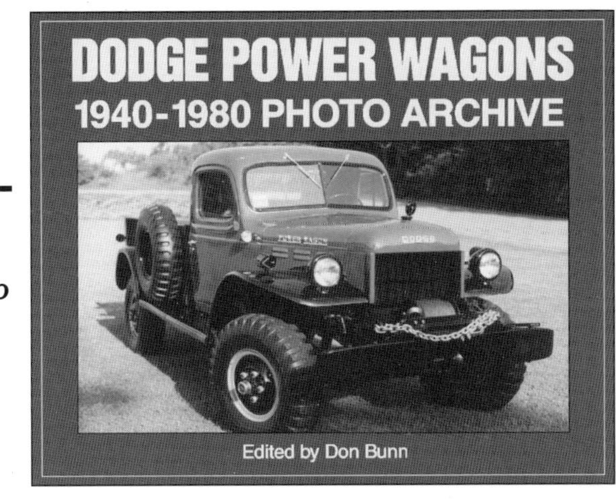

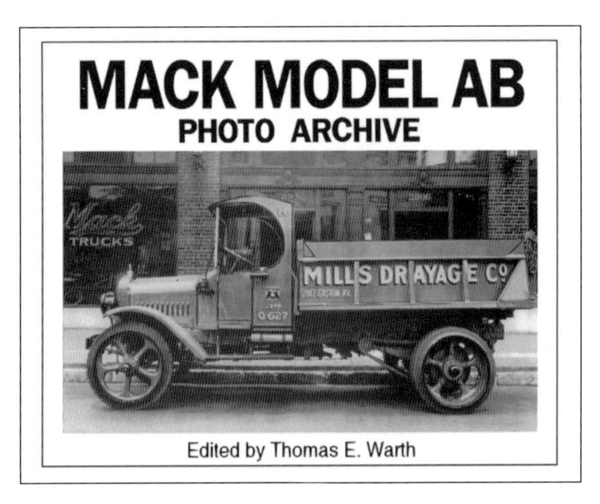